Tell me about...

PLANET EARTH

Written by
Emily Dodd

Illustrated by
Chorkung

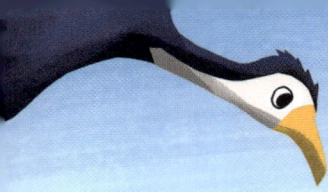

Contents

Earth is Home 8

Air 10

Light and Land 12

Reason for the Seasons 14

Weather 16

Water Cycle 18

Oceans 20

Rivers 22

Coasts 24

Caves 26

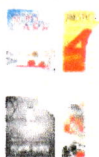

A TEMPLAR BOOK

First published in the UK in 2024 by Templar Books,
an imprint of Bonnier Books UK
4th Floor, Victoria House,
Bloomsbury Square, London WC1B 4DA
Owned by Bonnier Books
Sveavägen 56, Stockholm, Sweden
www.bonnierbooks.co.uk

1 3 5 7 9 10 8 6 4 2

ISBN 978-1-80078-345-4

This book was typeset in Catalina Clemente
The illustrations were created digitally

Edited by Ruth Symons and Rachael Roberts
Designed by Nathalie Eyraud
Production by Nick Read
Printed in China

Inside the Earth 28

Rock Recycling 30

Volcanoes 32

Earthquakes 34

Digging and Drilling 36

Energy 38

Climate Change 40

How You can Help
Planet Earth 42

Glossary 44

Earth is Home

We live on a brilliant ball of spinning rock called Earth. It's a planet travelling through space on a gigantic loop around a star called the Sun.

It takes a whole year to travel all the way around the Sun. So if you are five years old, you have circled the Sun five times already!

Earth travels around the Sun on an oval path but it also spins on the spot. The spin is why it gets dark at night.

There's another ball of rock about a quarter of the size of Earth and you can see it in the night sky. It's called the Moon.

It takes a month for the Moon to travel around Earth on an oval path.

Your home turns away from the Sun at night and by morning, it has turned back towards the Sun again. It takes 24 hours for a complete spin to happen, and we call that a whole day.

Air

Planet Earth is surrounded by air. You might think the air around you feels like it's nothing. But if you blow up a balloon, it is full of something, and that something is air!

The layer of air around Earth is called the atmosphere. It is made from a mixture of invisible floating things called gases.

One of those gases is water. That's right – water can be an invisible floating gas! It floats in the sky when it's a gas and then it turns back into a liquid and falls down as rain!

The atmosphere works like a window – it lets sunlight through to warm us up. But it protects Earth too. Dangerous light from the Sun hits the atmosphere and bounces off back into space.

Another important gas in our atmosphere is oxygen. Trees make oxygen and we breathe it in. It travels in our blood and it powers our body. All living things need oxygen to survive.

Light and Land

As the Sun's light shines onto Earth, it spreads out across Earth's round shape. If you shine a torch onto a ball, you'll see it do just the same thing.

The way the light spreads out across Earth gives us different temperatures and weathers. This changes how the land looks, too.

At Earth's top and bottom are the north and south poles. Sunlight is more spread out over these places so they're colder.

Tundra

Arctic

Desert

Savannah

Imagine slicing Earth into two equal halves. They would join at a middle line called the equator.

Places along the equator are hotter because lots of light shines down onto them all at once.

Swamp

Tropical rainforest

In between the poles and the equator are countries where sunlight is only a bit spread out. They have a medium temperature and four different weather patterns called seasons.

Temperate forest

Marsh

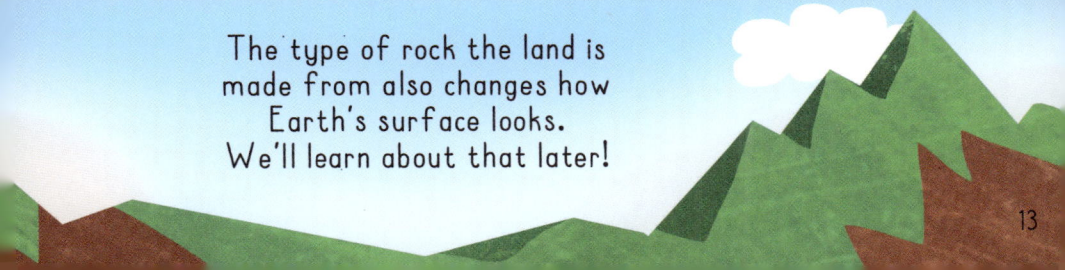

The type of rock the land is made from also changes how Earth's surface looks. We'll learn about that later!

Reason for the Seasons

The Earth is always spinning like a top, but it also leans to one side. The wonky spin is why we get different seasons throughout the year. Here's how it works...

When it is summer in the north half of Earth, it is winter in the south. This is because of how Earth leans.

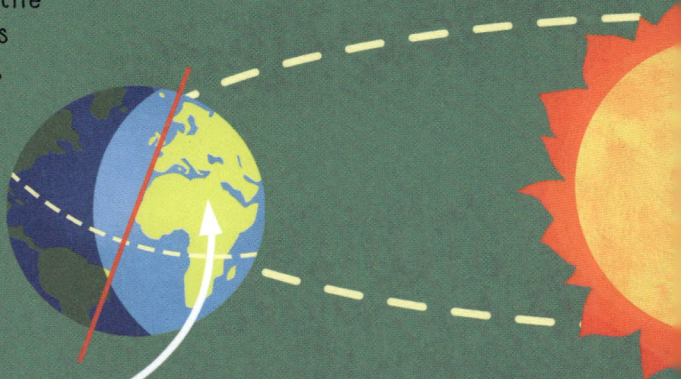

Earth leans towards the Sun for part of its year-long journey. That's when we get summer.

Summer
More light and warmth.

Plants and animals get used to the seasons and they grow and change in time with them.

There are two in-between seasons called spring and autumn. These happen when a country is tilted halfway away from the Sun and halfway towards it.

Earth leans away from the Sun later in the year. That's when we get winter.

Winter
Less light and warmth.

Did you know...?
A season is around three months long.

The north and south of Earth lean opposite ways, giving them opposite seasons.

Monsoons
Some countries only have two seasons: wet and dry. The wet season is called a monsoon. It rains a lot and it can last several months.

Weather

The Sun heats Earth unevenly so there are cold places and hot places. The hot air and the cold air move around to give us all kinds of weather.

Moving Air
Warm air rises up. Colder air sinks down. That's why a hot air balloon rises – it has hot air inside of it! Moving air becomes wind.

Currents
The same thing happens in the water. Warm water rises up and cold water sinks down. The water temperature changes the weather above the sea, too.

Wind that moves quickly across the ocean can turn into swirling storms called hurricanes.

Clouds

Rain clouds form when water gas in the atmosphere cools and turns into liquid. It falls as rain or freezes into snowflakes or hail.

Fog and mist are types of cloud that lie on the ground.

Lightning

Lightning is made in clouds when the wind moves tiny balls of water and gas about. The balls rub past each other and charge up with electricity. The electricity moves as lightning.

Water Cycle

The water on planet Earth moves around in a pattern called the water cycle. It changes from a liquid to a gas and then back to a liquid. Here's how it works...

Did you know...? Water changing from a liquid to a gas is called evaporation.

4. It blows along with the wind.

3. Water vapour rises into the air.

2. The very top layer of water escapes into the air. It turns into an invisible gas called water vapour.

1. The water in lakes and oceans is heated by the Sun.

Start here!

5. Water vapour cools down and makes shapes called clouds.

6. Water changes from a gas to a liquid and it falls as rain, hail or snow.

7. The rain flows down streams and rivers and back into the sea.

8. The whole thing starts all over again!

Did you know...? The water you drink is the same water the dinosaurs drank.

Oceans

If you flew out into space and looked back at Earth
it would look blue. That's because two thirds of our
planet's surface is covered in liquid water.
It's mostly found in the oceans and seas.

Waves
Waves are made on the
surface of the water as
the wind pushes the sea.

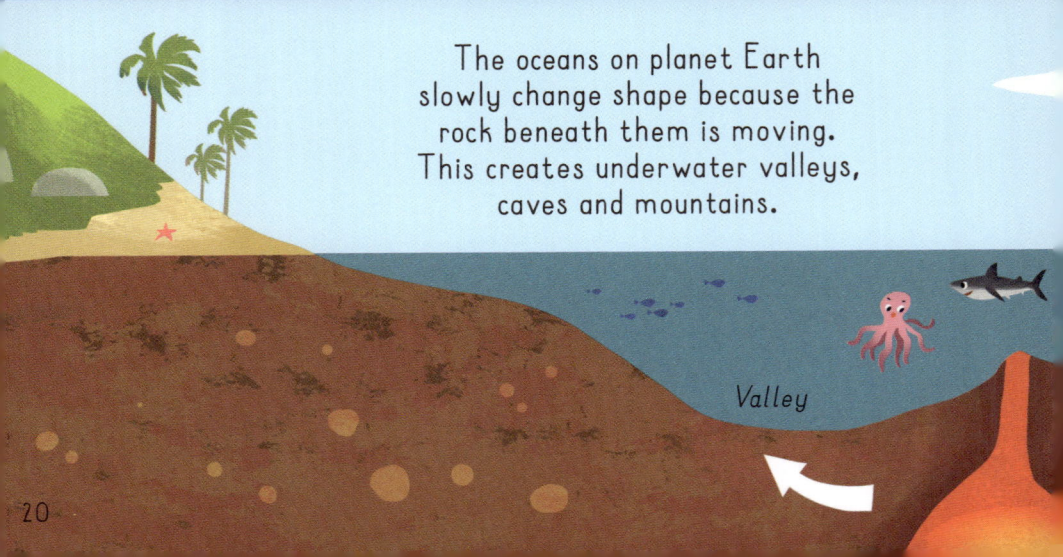

The oceans on planet Earth
slowly change shape because the
rock beneath them is moving.
This creates underwater valleys,
caves and mountains.

Valley

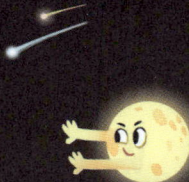

Tides

The sea comes in at high tide and goes out at low tide. This happens twice every day because of the way Earth is spinning beneath the Moon.

That's right, the Moon makes our tides! Gravity is a pull that happens between Earth and the Moon. It pulls on you too. When you jump, gravity pulls you back down to Earth.

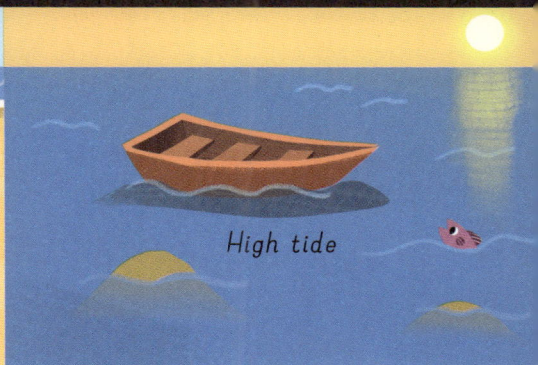

Low tide

High tide

Mountain

Did you know...?
Seawater is salty because of salt from rocks!

21

Rivers

Rivers aren't salty like oceans. The water that flows in rivers is called freshwater. Sometimes rivers begin underground. Other rivers fill up as ice melts or when rain falls.

If you put a ball on a slope, it rolls downhill. Rivers are the same, they flow downhill. It's all because of the pull of gravity.

Carving a channel

The river water rubs against the sides of the river and the river bed. This wears away the rock and soil as it flows.

Valleys

Over many years, rivers carve deep triangles into the land, called valleys.

Dropping off

When the river slows, it drops off the mud and rock it is carrying. This makes new land and changes the river's shape.

Carrying rocks

The river rolls rocks along and they bash into each other and break apart.

Waterfall

When a river travels over softer rock, it wears it down quicker than harder rock. Over time, this makes a step called a waterfall.

Coasts

The coast is the place where the land meets the sea. The waves and the weather wear away the coast and make it change shape.

Cliffs are steep walls of hard rock. They are hit by the weather at the top and bashed by waves at the bottom.

As the seawater crashes into the cliff, it breaks it away into caves and arches.

Cliff

Headlands
Hard rock takes longer to wear away, so it sticks out into the sea in pointy shapes called headlands.

Sand is made from tiny pieces of rock and shell. The rock gets broken down by the weather and carried to the beach by rivers. Shells wash up from the sea and get broken into sand by the waves.

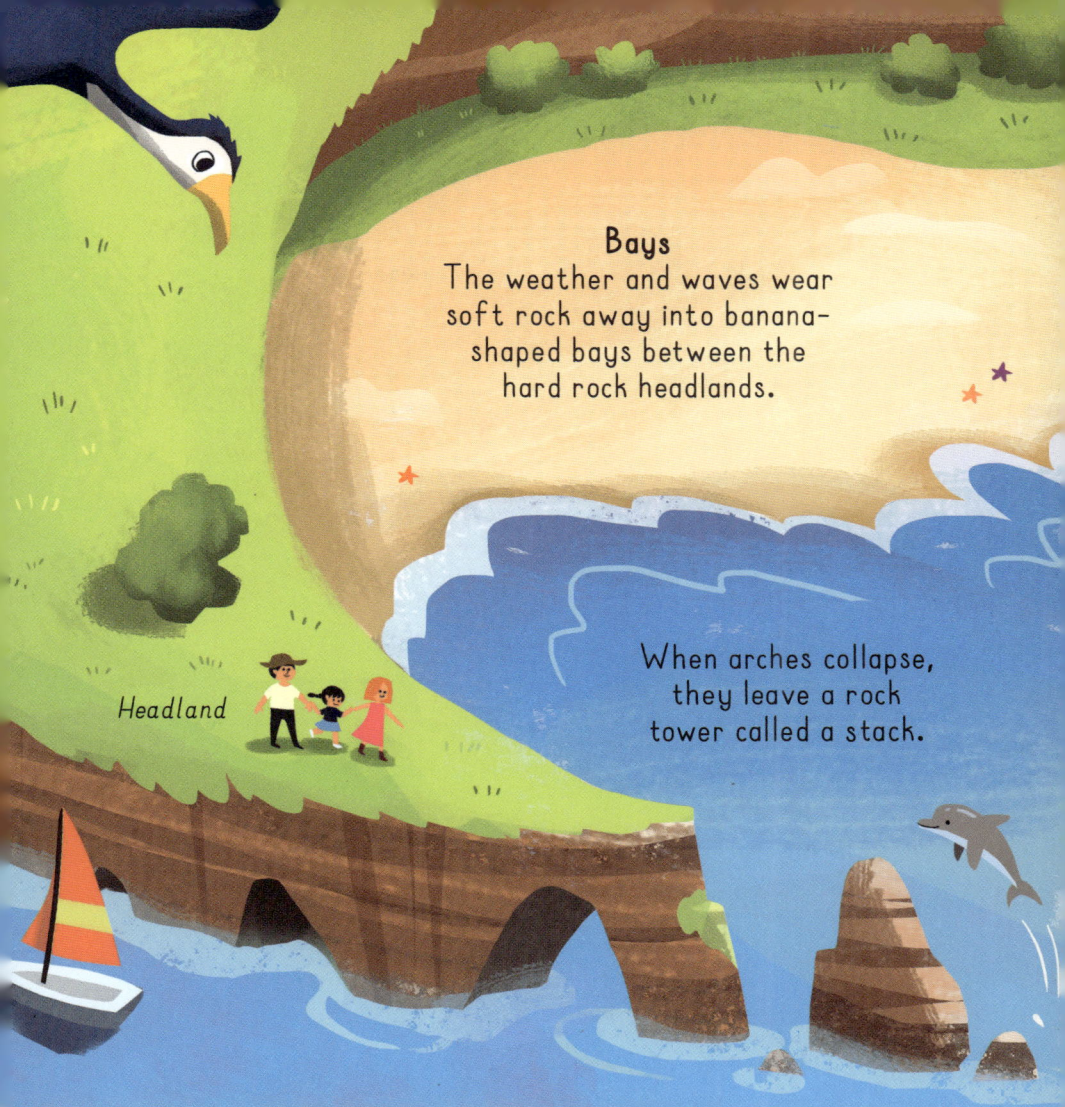

Bays
The weather and waves wear soft rock away into banana-shaped bays between the hard rock headlands.

Headland

When arches collapse, they leave a rock tower called a stack.

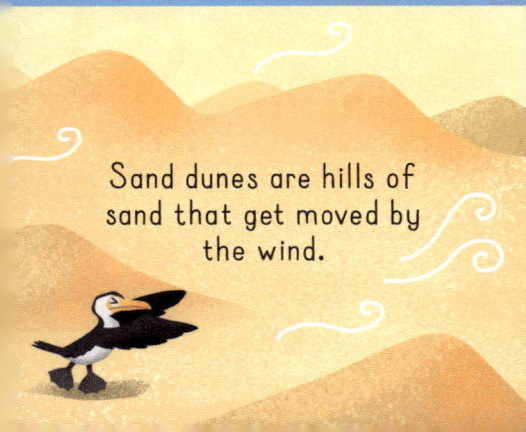

Sand dunes are hills of sand that get moved by the wind.

Did you know...?
Beaches made from pebbles are called shingle.

Caves

Caves are big holes carved into cliffs by waves hitting the rock. But they can also form underground as rain trickles through cracks in the rock.

That's right, tiny little rain droplets can make massive caves because they dissolve the rock away a little bit at a time.

Underground rivers flow through caves. They wear the floor of the cave away to make them even bigger.

Inside the cave, some droplets of rainwater evaporate. As the liquid water drops turn into gas, they leave behind the tiny bits of rock they were carrying. The bits of rock stick to the roof.

In a thousand years, all the drops of water will have left enough rock behind to make a shape about as long as your finger. This is called a stalactite.

The same thing happens as the water drips onto the floor of the cave too. The cave floor grows upwards into a wider pointy shape called a stalagmite.

Inside the Earth

If you could suck all the water away from planet Earth with a gigantic straw, you'd be left with a rocky looking ball. The rocky outer layer of planet Earth is called the crust.

If the whole Earth was the size of an apple, the crust would be the thickness of the apple's skin.

The crust is split into giant moving jigsaw pieces called plates. They move at the speed your fingernails grow – a few centimetres every year.

Beneath the crust is a thick layer of squashed hot rock called the mantle. The mantle flows around slowly, and the plates move around on top of it.

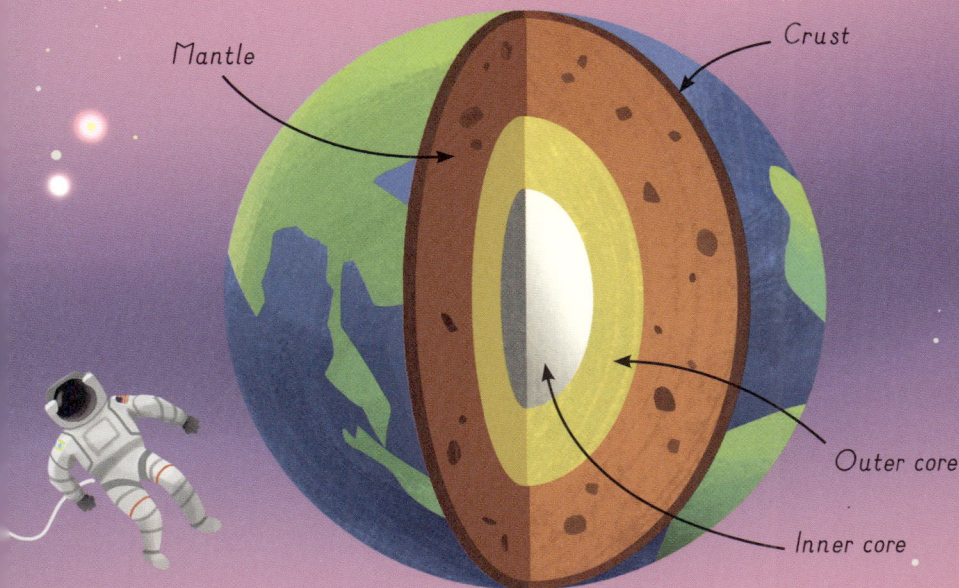

Mantle

Crust

Outer core

Inner core

Below the mantle is a hot metal ball called the core. Yes, there is metal in the middle of planet Earth, and this metal is even hotter than the surface of the Sun!

The inner core is made of solid metal and the outer core is made of liquid metal. The metal makes Earth a giant magnet!

Rock Recycling

Planet Earth recycles rock! It turns old rock into
new rock in a pattern called the rock cycle.
Here are some of the ways it happens...

Mountains are worn
down by water to
make mud and sand.

Sand sinks to the bottom of rivers,
lakes and the sea, and gets squashed
under layers of sand that fall on top.

When the seabed moves
towards the land, the thinner
and lighter ocean plate is pulled
down deep beneath the thicker
heavy land plate.

Sea water goes down the gap
too, and this makes the rock
spread out, rise up and melt.

Many years later, these
layers become a solid rock
called sedimentary rock.

Sometimes melted rocks cool inside volcanoes and form crystals inside. The rock made by volcanoes is called igneous rock.

The thick land plate crumples up to make mountains. The rocks underground move and sometimes come up to the surface.

Rocks get squashed deep below the ground too. They change shape without melting and become metamorphic rock.

Volcanoes

Mountains with hot melted rock and gas inside them
can explode or erupt. We call these mountains volcanoes.
When they erupt, the mountain shatters into tiny
pieces called volcanic ash.

There are other ways a volcano
can erupt. Sometimes melted rock
called lava flows from a volcano in
rivers. Other volcanoes shoot out
hot lava balls called lava bombs
or a bubbly rock that floats,
called pumice.

Some volcanoes are
underwater cracks
where two plates pull
apart. Lava comes out
and cools to make new
land in the middle.

Did you know...?
When lava cools really quickly it can make glass!

Hawaii

Volcanoes can even make new islands!
When melted rock punches up through
the rocky sea floor, lava flows out and
cools to make an underwater hill of solid
rock. More hot rock punches through
the hill and cools to make it even bigger.
In time, an island will pop out of the sea,
just like Hawaii did.

Earthquakes

The plates that make up Earth's crust are slowly moving around on the surface of our planet. But if a plate moves suddenly, it makes the ground shake in an earthquake.

When the plates try to slide past each other, they bend and get stuck. When they finally spring back, the sudden slip and slide of rock makes the earth shake.

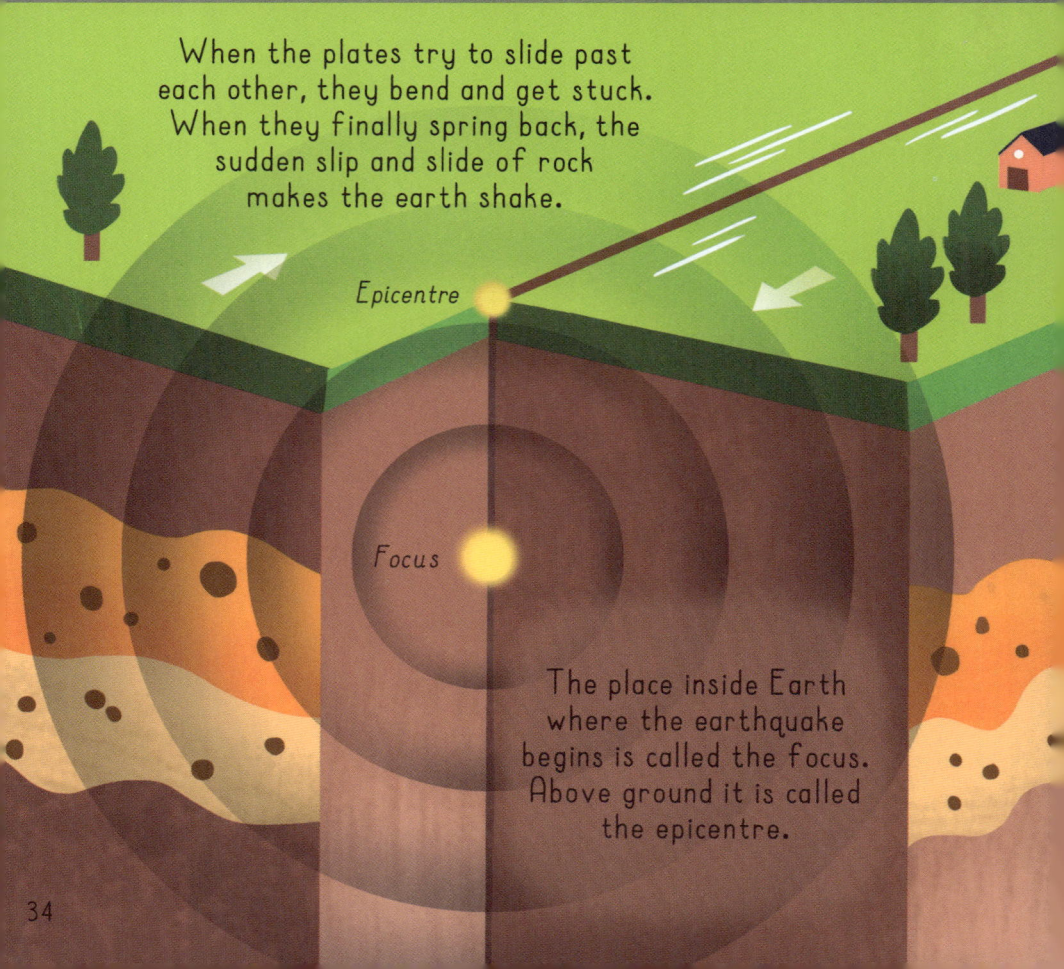

Epicentre

Focus

The place inside Earth where the earthquake begins is called the focus. Above ground it is called the epicentre.

Earthquakes travel out from the focus in waves. They travel inside the planet and across the surface. Even the rocky plates move in a wave!

Seismometer

We measure earthquake shakes with an instrument called a seismometer. The earthquake shakes a pen that draws a squiggly line to show us how big the earthquake shakes are.

Tsunami

When rock in the seabed moves suddenly, it can move the water above. The moving water piles up into a big wave, and when it reaches the land, it's called a tsunami.

Digging and Drilling

When humans dig useful rocks and metals out of the ground, it is called mining. People also drill long holes deep down into the rock to find little pockets of gas and a liquid called oil.

The oil and gas found deep underground were once tiny sea creatures. They sank to the bottom of the sea and got squashed over millions of years. They turned into a dark liquid called oil and a gas called methane.

Coal is a black rock that gives off lots of heat when it burns. It is made from leaves that sunk in swamps millions of years ago.

We can burn oil, coal and methane gas to make electricity and to power vehicles.

Metals can make lots of useful things including bikes, phones, computers and cars.

Metal
Most metals are hidden underground within other rocks. A few metals are found just as they are at the surface, including gold, silver and copper.

Energy

The Sun gives out heat and light energy. It helps us to see and it keeps us warm. We can make our own heat and light energy too when we burn things, or if we use electricity.

Electricity is a type of energy that is used in our homes to make lights and devices work. Cars can charge up using electricity too.

When we burn coal, oil and gas in a power station, the fire heats up a pipe of flowing water and that water turns into boiling hot steam.

The steam makes a big fan spin around, and that makes electricity!

Coal, oil and gas come from inside Earth's crust and one day they will run out. But we can make electricity in other ways that don't run out:

The wind spinning a wind turbine.

Water trapped by a dam flowing over a water wheel.

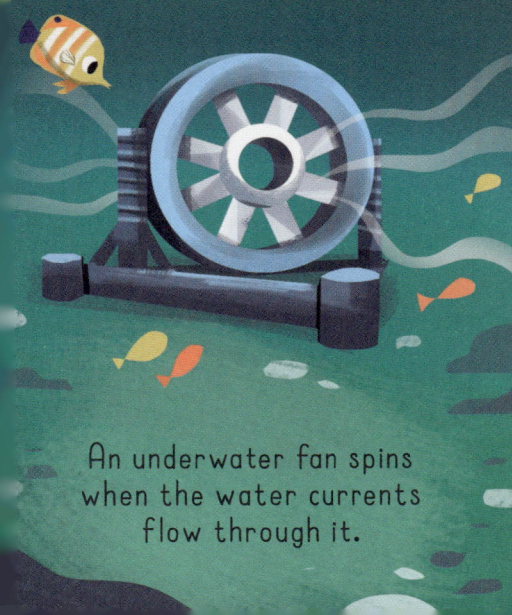

An underwater fan spins when the water currents flow through it.

Catching the Sun's light and heat with solar panels.

39

Climate Change

Carbon is something stored inside all living things. When we burn trees, the carbon inside them escapes into the air as a gas called carbon dioxide. The same happens when we burn coal, oil and gas.

Carbon dioxide acts a bit like a blanket by trapping heat. The gas causes sunlight to reflect back down towards Earth. The trapped sunlight heats up our planet. This is called global warming.

Humans have been burning things in power stations for 300 years! That has let lots more carbon dioxide into our atmosphere so more sunlight is getting trapped.

As Earth gets warmer it makes the patterns of seasons change and we get surprise weather. This is called climate change.

If we keep burning things, the planet could heat up so much that the ice on land melts. If this happens, the extra water will flow into the sea to make it rise. Then the animals and people living on low down land would be flooded.

How You can Help Planet Earth

Climate change is harming the homes of plants, animals and humans. But there are things you can do to help Earth.

Use less electricity

Switch off lights when you leave a room and turn off electric devices when you're not using them.

Most electricity is made in ways that let carbon dioxide into the air. The less electricity we use, the less gas goes into the atmosphere to warm our planet.

Recycle and reuse

Don't throw things away if you can recycle or reuse them! Using things again or using old materials to make new things uses less electricity than making new things from scratch.

Use less water
Cleaning water and
pumping it to our
homes uses electricity.

Plant trees
Trees clean the air by
removing carbon dioxide from
it and releasing oxygen gas.

Green power
Make electricity from
wind, waves and the Sun.

Green transport
Walk, cycle or take public
transport instead of using a
car, when you can.

Glossary

Atmosphere
The layer of air surrounding planet Earth. It is made up of a mixture of gases.

Carbon
A material found in rocks and in all living things. When these things are burned, the carbon inside them becomes a gas. It joins with the oxygen in our air to become carbon dioxide.

Climate Change
A change in weather and seasons over time. This includes warmer and wetter weather, and more extreme weather. It happens because of global warming.

Current
A flow of moving water in a certain direction, for example, water moving from a hot area to a cold area.

Energy
Everything that does something has energy. It is the power to get things done. Light, heat, electricity, sound and movement are types of energy.

Equator
An imaginary line around the middle of Earth where the Sun's heat and light is least spread out over the planet's surface. This makes it the hottest part of Earth.

Evaporation
When a liquid turns into a gas.

Gas
A material made up of tiny pieces that spread out and are so small you can't see them. A gas doesn't have a solid surface and it spreads out to fill a container, like air in an empty bottle.

Gravity
An invisible force that pulls things towards each other. Everything has gravity, but bigger objects have more than smaller ones. Earth's gravity pulls on the Moon and the Moon's gravity pulls our tides.

Liquid
A material made up of bits that slide past each other but that still stick together. A liquid flows and can be poured, for example, rain water.

Oxygen
A gas in our air that all living things need to survive. We breath it in and it helps our bodies work.

Season
A weather pattern that happens for part of the year, like summer or winter.

Temperature
A measure of how hot or cold something is.

Tide
The rising and falling of the sea twice a day because of the pull of the Moon's gravity. At high tide, the water comes higher up the beach. At low tide the water goes back down again.

Valley
A V or U-shaped dip in the land. Valleys can be made when flowing water wears away the land.

Water Vapour
The name for water when it is a gas. It is invisible and floats in the air. When water vapour cools in clouds, it changes into a liquid or solid and falls as rain, snow or hail.